# S PEECH SIGNAL
## PROCESSING

By Prof. Vishnu Narayan Saxena

Any audio signal or music signal contain a range of frequencies from 20 Hz to 20 Khz . In this frequency range band of music signal there are several sub bands of different frequencies where every frequency band has its own characteristics and sound effect. Every band of frequencies having a different information about music signal. And according to our need and requirement of any particular application we do require to amplify or to attenuate any particular frequency band . In this book my whole effort is to give readers enough practical knowledge that they can understand the difference between different frequency components and can separate, amplify and attenuate desired frequency components.

Prof. V.N. Saxena

Department of technical
education and skill
developement

INDIA

Vishnunarayan.saxena@gmail.com

13/03/2016

# DEDICATION

THIS BOOK IS DEDICATED TO MY FATHER MR. B. N. SAXENA BY WHOM I LEARNED THE LESSION OF HARD WORK, CONSISTENCY AND LEARNING ATTITUDE.

# ABOUT THE AUTHOR

Prof Vishnu Narayan Saxena is currently working with U.P.C. an AICTE approved Institution under the department of technical education (D.T.E.). Govt. of Madhya Pradesh India.

He has received his Master of engineering in communication control and networking from M.I.T.S. Gwalior and his Bachelor of engineering from I.T.M. Gwalior

Prof Saxena has been in field of education and research science 2003. He has been published many International/national research papers on different field of an engineering and has been published many e-books with different repudiated publications

His many articles has been published and available online at Scribed, Lulu publication, Slide share, Amazon

He has been participated in many FDP organized by different IITs (Indian institute of technology) and NITs (National institute of technology) and completed many online course (ALISON)

His field of research and interest are Digital signal processing, Digital Image Processing, Audio signal processing, Wavelets and filter design.

# ACKNOWLEDGEMENT

I want to dedicated this work to my father Mr B.N. Saxena by whom hard working capacity and consistency has been transferred into me. My wife Pooja who always appreciate me to do good job and for ignoring all of my mistakes committed by me knowingly or unknowingly and my cute daughter Poorvi who developed into me a sense of responsibility and patience.

I am very thankful to my wife and co author of this book prof. pooja saxena .Who discussed to me about every topic of this book and help me to improve our work always she brings to me with new ideas and suggestions.

I am very grateful for every person by whom I could learn. I am also grateful for all the situation and incidences by which I learn something Again I want to say thanks technology (Google search engine and internet) which help me a lot to understand this subject and help me to share my views to others. I am also thankful to all the authors and writers who share their research work with us.

As an Author of this book I will try to arrange all the topics and subtopics in systematic way and in proper sequence. Because as teacher I always felt that only knowledge is not sufficient but more important thing is that how we can transferring our knowledge to other. If things are not arranged in systematic form and in proper sequence then the things becomes very hard whereas if we can arrange the things in systematic form and in proper sequence then the same things becomes easy. When any Subject is known to us then suddenly we found that it becomes very easy whereas when any subject is not known to us it seems very tough. My whole effort behind writing this book is that how to make technology approachable to the young brain. How to make technology easy and interesting.

Again as we know about any subject and when we developed an understanding about any subject we suddenly found that how to approach that subject. Of course book writing is a very beautiful solution of approaching that subject or topic

**AUDIO SIGNAL PROCESSING:** Audio signal for example a music signal contains a frequency bandwidth from 20 Hz to 20 KHz. In this bandwidth there a many sub bands which has different sound. As we know a music signal is composed with the help of many music instruments for example Drum, Guitar, Violin, Harmonium etc the sound of different instruments have different characteristics and cover different audio frequency range. How we can separate these different frequency bands from each other and how we can attenuate or amplify any particular frequency band is the main object of this book. In our home appliances for example in audio player or radio there is option of Bass Treble and volume when we increase the bass then the low frequency components (for example the sound of drum )of music signal gets increase when we increase the treble the high frequency components (for example the sound of guitar) of any music signal increases.

## SIGNAL:

Any signal in time domain is considered as raw signal. The Propose of all Transformation techniques are to convert time domain signal in a form so that desired information can be extracted from these signal's and after the application of certain transform the resultant signal is known as processed signal.

## TIME DOMAIN ANALYSIS OF SIGANL :

In time domain representation of the signal there is a graph between time and amplitude. The time domain representation of the signal gives information about the signal that at which time instants what is the amplitude of the signal or how amplitude of the signal is varying with respect to time but it gives no information about the Different frequency contents that are presents in the signal.

## FREQUENCY DOMAIN ANALYSIS OF SIGNAL:

In frequency domain representation of the signal there is a graph between frequency and amplitude .The frequency domain representation of the signal tells us that in a given signal what different frequency components are present and what respective amplitudes are there. But again frequency domain representation gives no idea that at which time these frequency components are presents. In some applications frequency domain representation is more important. Or we can say that frequency domain representation gives more information about any signal (for example audio music signal).

**TIME FREQUENCY ANALYSIS OF MUSIC SIGNAL.**
Time domain representation of the signal gives no information about the Different frequency contents that are presents in the signal. frequency domain representation gives no idea that at which time different frequency components are presents in any signal. So time frequency presentation of any signal gives an idea that at which time what different frequencies are present in any signal

**DIFFERENT TRANSFORMATION METHODS :**

There are different transformation methods or techniques which are used for transforming signal from one domain to an another domain

- **DISCREET FOURIER TRANSFORM**
- **SHORT TIME FOURIER TRANSFORM**
- **WAVELET TRANSFORM**

**1. DISCRETE FOURIER TRANSFORM:**

The Discrete Fourier transform is used for converting time domain signal into frequency domain

Suppose

x= 100*sin (2*pi*2*t) +50*cos (2*pi*3*t)

 is a time Domain representation of a given signal and

y= fft (x)

 is frequency domain representation of signal x

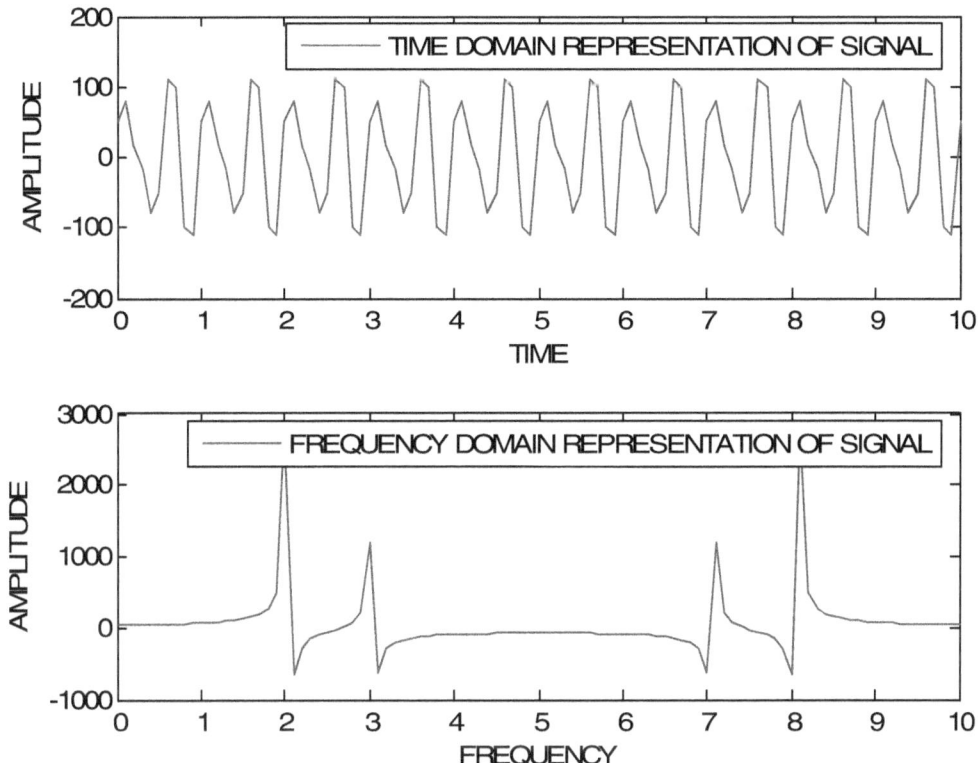

**Fig Time Domain and frequency Domain representation of signal**

As it is clear from fig that time domain representation of signal gives no idea about the frequency components of signal it simply shows that how the amplitude of the signal is varying with respect to time whereas frequency domain representation of the signal shows the different frequency components presents in a signal and their respective amplitudes but it gives no idea that at which time instants these frequency components presents.

In the other word we can say in time domain representation of signal the time resolution of signal is very high but its frequency resolution is zero because it gives no idea about different frequency components presents in a signal where as in frequency domain representation of the signal the frequency resolution of the signal is very high but its time resolution is zero because it gives no idea about time.

Both domain of signal analysis has its own utilities and has its own importance and having its own sets of advantages and disadvantages.

How Fourier transform convert time domain signal into frequency domain. Suppose x (t) shows the time domain representation of signal and X (f) shows the frequency domain signal Then

$$X(f) = \int_{-\infty}^{+\infty} x(t).e^{-2j\pi ft} dt \qquad\qquad \text{..} \qquad (1)$$

$$x(t) = \int_{-\infty}^{+\infty} X(f).e^{2j\pi ft} df \qquad\qquad \text{.} \qquad (2)$$

Where

$$e^{2j\pi ft} = Cos(2\pi ft) + j \; Sin(2\pi ft) \qquad\qquad (3)$$

Equation (1) and equation (2) gives Fourier transform and inverse Fourier transform of any signal the exponential term of equation (1) can be expressed in terms of sine and cosine function shown by equation (3)

With the help of equation (1) and equation (2) we can convert any time domain signal into frequency domain signal and frequency domain signal into time domain signal respectively.

As clear from equation (1) that in Fourier transform the signal is integrated from – infinity to + infinity over time for each frequency In the other word we can say that equation 1 take a frequency for example $f_1$ and search it from –infinity to + infinity over time if it find the $f_1$ frequency components it simply adds the magnitude of all f1 frequency components.

Again take an another frequency for example $f_2$ and search it from – infinity to + infinity over time if it find the $f_2$ frequency components it simply adds the magnitude of all f2 frequency components

Again repeat the same process with f3,$f_4$, $f_5$……. and so on

No matter in time axis where these frequency components exits from – infinity to + infinity it will affect the result of integration in the same way .

For every frequency Fourier transform check that whether this particular frequency component present or not present in time from minus infinite to plus infinite. And if present then how many time these particular frequency components presents and what is the amplitude of this particular frequency component and then simply add that particular frequency component and calculate the amplitude of any particular frequency component.

Again take a second frequency component and check that in time from minus infinite to plus infinite how many times this particular frequency component exists and what is amplitude of this particular frequency component and then simply adds them. In this way the Fourier transform calculated the amplitude of every frequency components presents in a given signal and draw a graph between frequency and amplitude

Again there are certain disadvantage of frequency domain representation of the signal first disadvantage is that it gives no idea about time .The frequency transform of any signal simply tells us that in any given signal what spectral components are present and what are their respective amplitudes but it gives no idea that in time axis where these frequency components exists

So again the D.F.T. prove its suitability for the signals which are stationary in nature but this transform is not suitable for non stationary signal

By stationary signal we simply means the signal in which the frequency does not change with respect to time or we can say that all frequency components exits for all the time

By non stationary signal we simply mean the signal in which frequency changes with respect to the time. or in which all the frequency components does not exist for all the time interval .but some frequencies are exits for some particular time interval whereas some other frequency components exists for some other time interval.

To understand the suitability of the DFT only for stationary signal and not for non stationary signal take the following example suppose there are two signals $S_1$ and $S_2$ the signal $S_1$ has three frequency components $f_1, f_2$ and $f_3$ all the times and suppose the signal $S_2$ contains the same frequency components f1,f2 and f3 but for the different -different time interval's so we can say that these two signals are completely different in nature. But in spite of it the Fourier transform of these two signals will be the same because these two signals have the same frequency components of course one signal contains all the frequency components all the time whereas second signal contains these frequency components at different time intervals but as we know that Fourier transform has nothing to do with the time. No matter where these frequency components exits over time the matter is only that whether they occur or not and what are their amplitudes.

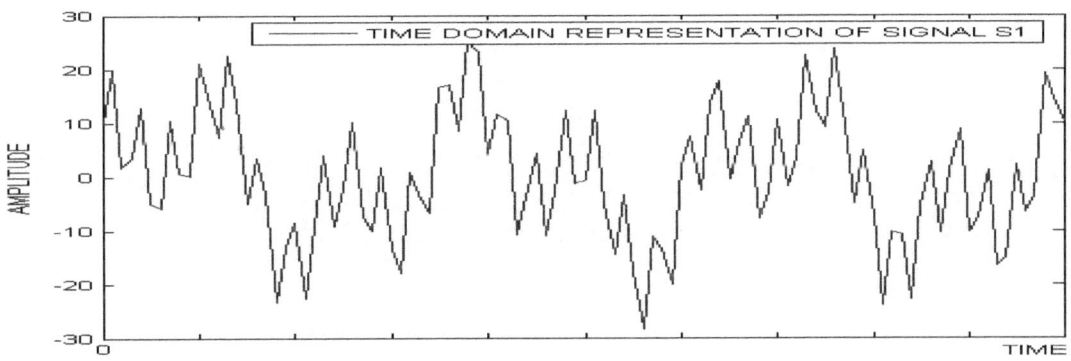

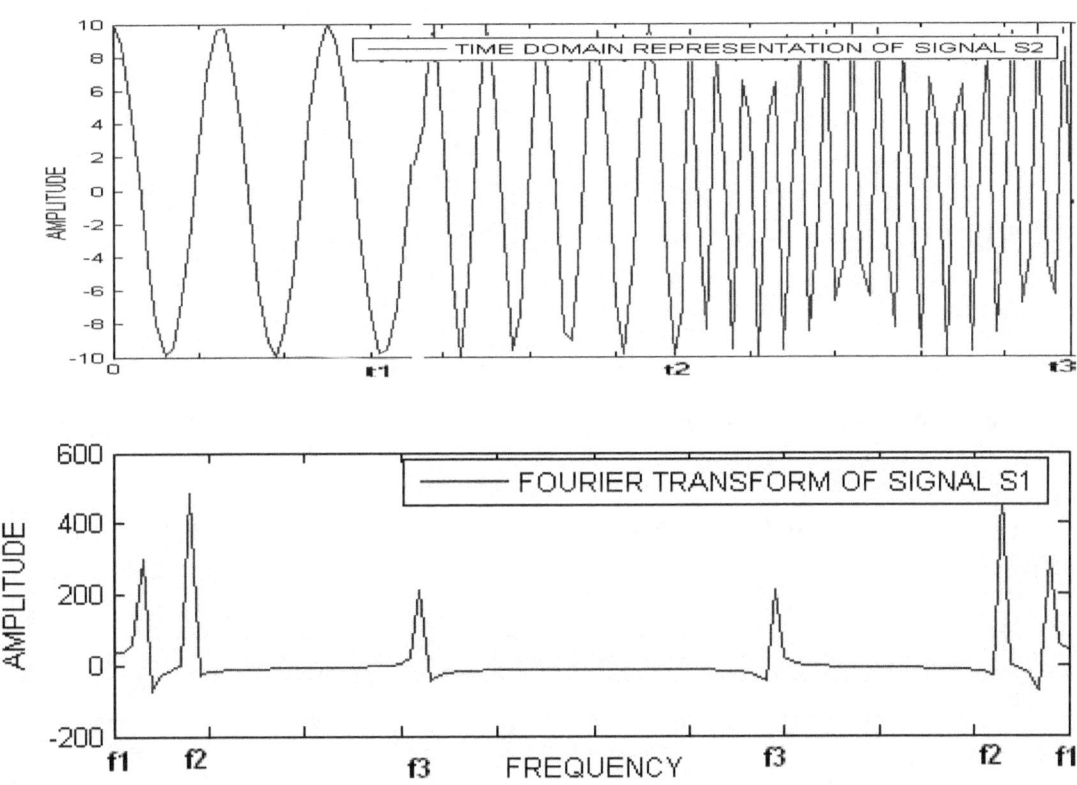

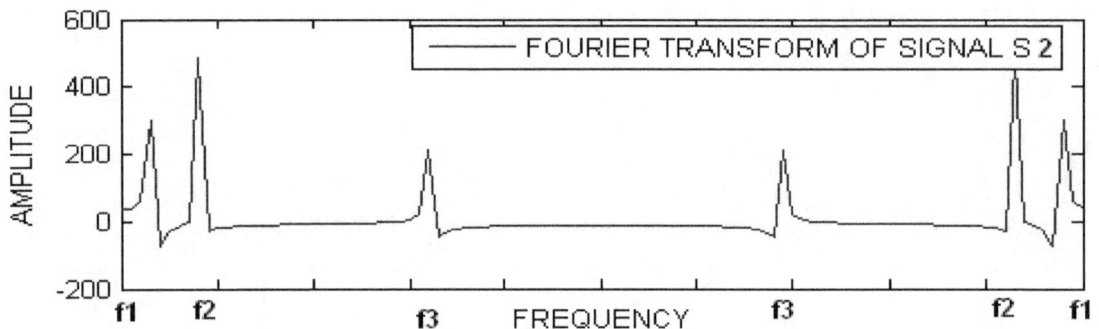

**Fig   Time domain and frequency domain representation of stationary and non stationary signals**

Again take an another example suppose there are two signals $S_3$ and $S_4$. **S**ignal $S_3$ contains frequencies $f_1$ for time interval $t_0$ to $t_1$, frequency $f_2$ for time interval $t_1$ to $t_2$, and frequency $f_3$ for time interval $t_2$ to $t_3$. **A**nd signal $S_4$ contain frequency $f_3$ for time interval $t_0$ to $t_1$, frequency $f_2$ for $t_1$ to $t_2$ and frequency $f_1$ for time interval $t_2$ to $t_3$

So we can say that these two signals are quiet different though both signals are having the same frequency components in same amount but the time instances where these frequency components exists are different so the overall characteristics of above these two signals S3 and S4 will be different but in spite of this the Fourier transform of these two signals will be the same because Fourier transform has nothing to do with time.

Fourier transform simply watch that what frequency components any signal has and what are their respective amplitudes

## 2. Short time Fourier transforms

The Short time Fourier Transform is a modified version of Fourier transform. S.T.F.T.  is nothing it is simply the Fourier transform of any signal multiplied by a window function.

$$STFT_X^{(w)} (t, f) = \int_t [x(t) . w^*( t - t')] . e^{-j2\Pi f t} \; dt \; ..............................4$$

The basic idea behind the STFT is that any non stationary signal can be considered stationary for a short time interval. So we can say that the STFT gives an idea about time frequency and amplitude.

But again the problem with STFT is that how to choose the size of window (time interval of window) because .Any single window size is not suitable for the analysis of all frequency components small window size is suitable for the analysis of high frequency components whereas the large window size is suitable for the small frequency component.

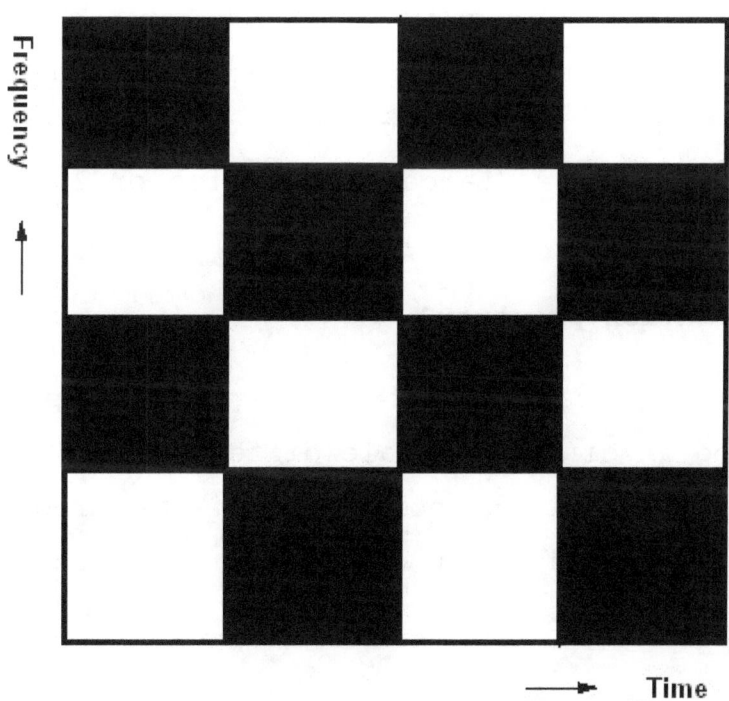

**Fig (3) STFT has same window size for all frequency component**

But the problem with STFT is that the window size remain same for the all analysis we can't choose different -different window Size for the analysis of different frequency components .As shown in fig that for all frequencies the size of window is same

Hence we can say that there is some resolution problem with S.T.F.T. at the same time we can't obtain both time as well as frequency resolution means we cannot know Exactly at which particular time instant which particular frequency components   Exists   We   can   know   only   that   in   which   time intervals(not  time  instants)  what  frequency  spectrum  occurs(not  exact frequency)

## 3. Wavelet transform

$$\varphi_x^\varphi(\tau,s) = \frac{1}{\sqrt{|s|}} \int_{-\infty}^{+\infty} x(t) \; \varphi^*\; (t \; - \; \tau)/s \; dt$$

........................  **(5)**

By  using  wavelet  Transform  we  can  overcome  the  problem  with  S.T.F.T.  in wavelet  transform  we  used  different  window  size  for  different  frequency components.  Low  scale  (small  window  size  or  small  time  scale)  is  used  for high  frequencies  and  high  scale  (Large  Window  size  or  large  time  scale)  is used for low frequencies

## 3.1 What is Multi resolution property?

Multi resolution  property means different frequency components presents in any signal are resolved at different scale(different time scale or different window size).Scale  is  inversely  proportional  to  the  frequency  means  small scale  is  used  for  higher  frequencies  whereas  large  scale  is  used  for  small frequencies

Small  scale  (Small  window  Size  or  Small  time  scale)is  used  for  the  analysis of higher frequency components. Means wavelet transform provide higher time resolution  for  high  frequency  means  if  any  signal  contain  a  very  high frequency  component  then  with  the  help  of  wavelet  transform  we  can  know  that at  which  exact  time  interval  (very  small  time  interval)  these  frequency components exists.

whereas the large scale (large window size or large time scale)is used for the analysis of small frequency components .Wavelet transform provide good time resolution for higher frequencies whereas for small frequencies time resolution is not so good .But if we study real word signals then we find that generally higher frequencies are occurs only for very short time interval whereas small frequency components are presents for long time interval. So wavelet transform prove its suitability for real word signals.

## 3.2 What is spectral localization?

Spectral localization property means that wavelet transform tells us that what frequency components are present in any given signal and at time axis where these frequency components are presents

## 3.3 What is Scale?

Scale is inversely proportional to the frequency. Large scale is used for the analysis of small frequency components presents in any signal Whereas Small scale is used for the analysis of high frequency components of any signal

Fig (4) DWT Uses different scale for different frequency components

## 1 READING A MUSIC SIGNAL.

```
N= 240000;
[x,fs,bits]=wavread('C:\Program Files\MATLAB\R2013a\bin\Bass\signal1.wav',N);
N,    240000
bits, 16,
fs,   44100,
```

Waveread is a Matlab function which is used to read an audio file (file with .Wav extension). In above function.
X = is the variable where audio samples store.
N = is the number of samples of an audio file here we read two lakh forty thousand samples (240000).
Bits = is a variable which stores number of bits to represent every sample in this example 16 samples are used to represent every sample.
fs = is a variable which stores the number of samples played per second in current example we are playing 44100 samples per second.

## 2. PLAYING A MUSIC SIGNAL AT DIFFERENT RATES.
### 2.1 : PLAYING AN AUDIO SIGNAL AT ITS ORIGINAL RATE
```
N= 240000;
[x,fs,bits]=wavread('C:\Program Files\MATLAB\R2013a\bin\Bass\signal1.wav',N);
wavplay(x,fs);
pause;
```

In above matlab example wavplay is the matlab function which is used to play back audio file (.wev extension audio files) at a rate fs(44100 ) samples per second

**PLOTING AN ORIGINAL AUDIO SIGNAL**
plot(x)

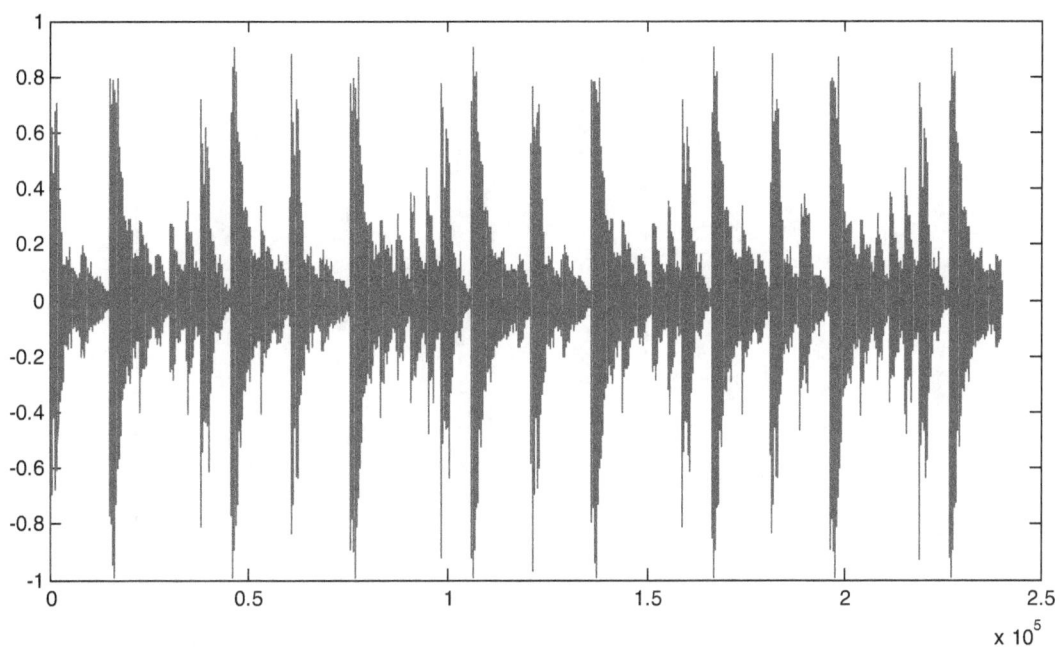

**2.2 : PLAYING AN AUDIO SIGNAL AT TWICE RATE**
N= 240000;

[x,fs,bits]=wavread('C:\Program Files\MATLAB\R2013a\bin\Bass\signal1.wav',N);

wavplay(x,2*fs);

pause;

In above matlab example wavplay is the matlab function which is used to play back audio file (.wev extension audio files) at a rate twice of fs(88200 ) samples per second.

```
>> N= 240000;
>> [x,fs,bits]=wavread('C:\Program Files\MATLAB\R2013a\bin\Bass\signal1.wav',N);
>> wavwrite(x,2*fs,bits,'C:\Program Files\MATLAB\R2013a\bin\Bass\twicerate.wav');
>> [y,fs,bits]=wavread('C:\Program Files\MATLAB\R2013a\bin\Bass\twicerate',N);
>>  wavplay(y,fs);
```

Wavwrite is a matlab commond which is used to write (store) wav file at any location on the disk in wabwrite function we also define the number of samples per second and number of bits used to represent every sample.

**PLOTTING TWICE RATE SIGNAL**
```
plot(y'r')
```

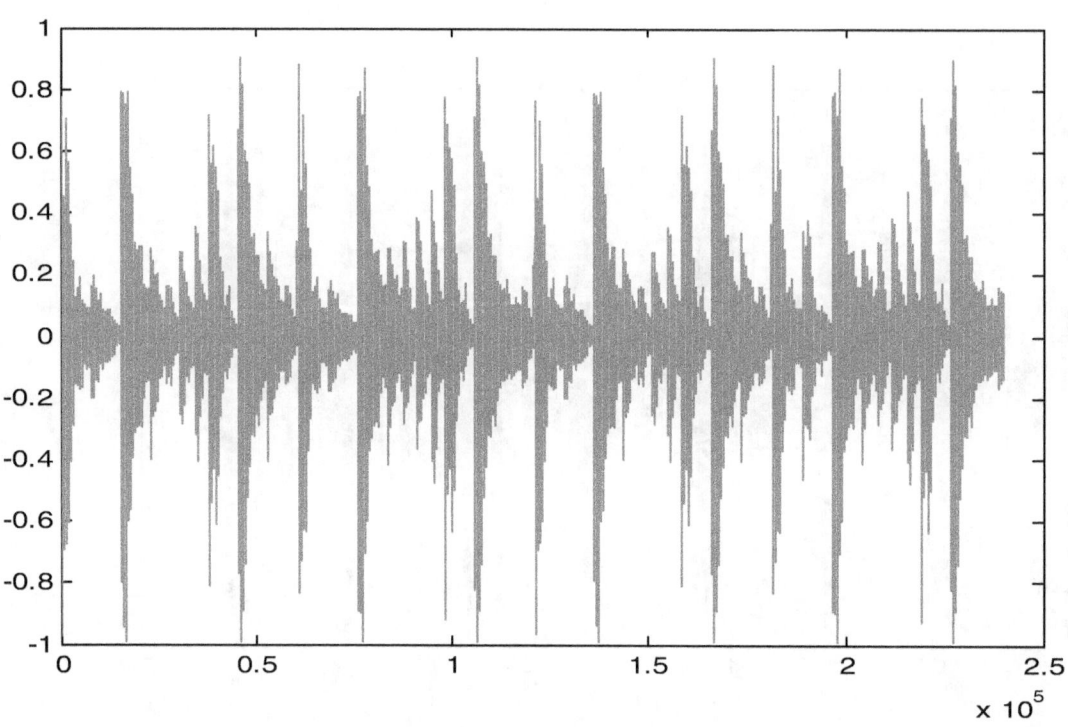

**2.3 : PLAYING AN AUDIO SIGNAL AT HALF RATE**
```
N= 240000;

[x,fs,bits]=wavread('C:\Program Files\MATLAB\R2013a\bin\Bass\signal1.wav',N);

wavplay(x,fs/2);

pause;
```

```
>> N= 240000;
>> [x,fs,bits]=wavread('C:\Program Files\MATLAB\R2013a\bin\Bass\signal1.wav',N);
>> wavwrite(x,fs/2,bits,'C:\Program Files\MATLAB\R2013a\bin\Bass\halfrate.wav');
>> [z,fs,bits]=wavread('C:\Program Files\MATLAB\R2013a\bin\Bass\halfrate.wav',N);
>> wavplay(z,fs);
```

In above matlab example wavplay is the matlab function which is used to play back audio file (.wev extension audio files) at a  rate half of fs( 22050) samples per second

**PLOTING HALF RATE SIGNAL:**

```
plot(z,'g')
```

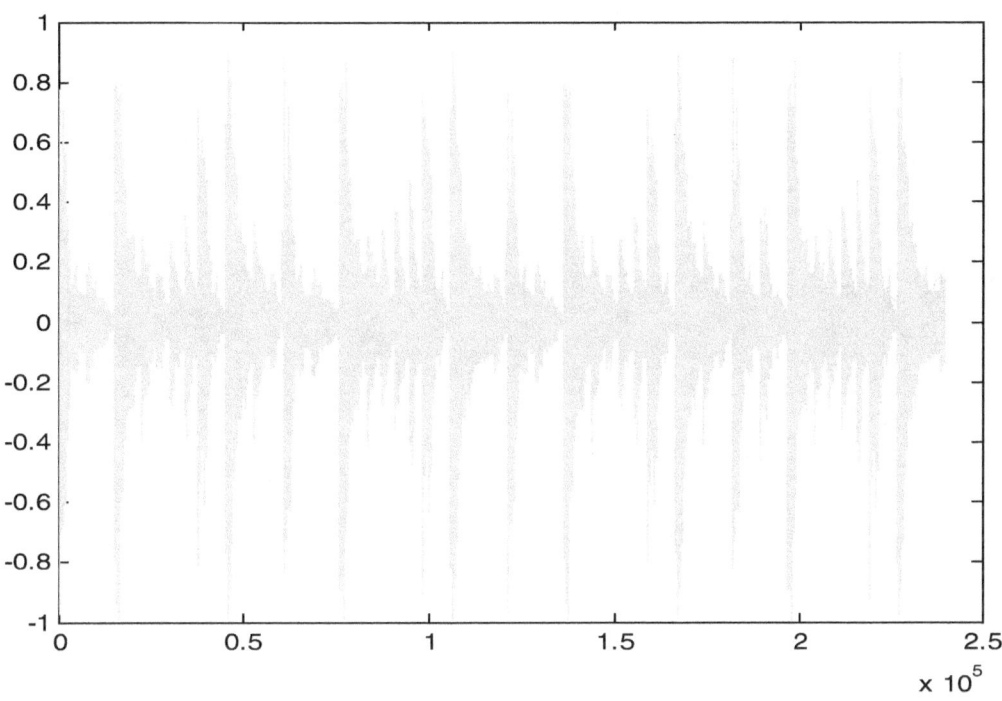

**CHANGING THE VOLUME OF A MUSIC SIGNAL**

**TWICE THE VOLUME OF MUSIC SIGNAL**

```
N= 240000;
[x,fs,bits]=wavread('C:\Program Files\MATLAB\R2013a\bin\Bass\signal1.wav',N);

wavplay(2*x,fs);
pause;

N= 240000;
[x,fs,bits]=wavread('C:\Program Files\MATLAB\R2013a\bin\Bass\signal1.wav',N);
wavwrite(2*x,fs,bits,'C:\Program Files\MATLAB\R2013a\bin\Bass\twicevolume.wav');
[TVOL,fs,bits]=wavread('C:\Program Files\MATLAB\R2013a\bin\Bass\twicevolume.wav',N);
wavplay(TVOL,fs);
```

Wavwrite function is used to write .wav extension file to the computer Memory .As we multiply variable x by 2 which result volume is increased by two times. fs is sampling rate. And bits shows the number of bits per sample.

PLOTING TWICE VOLUME SIGNAL
plot(TVOL,'r')

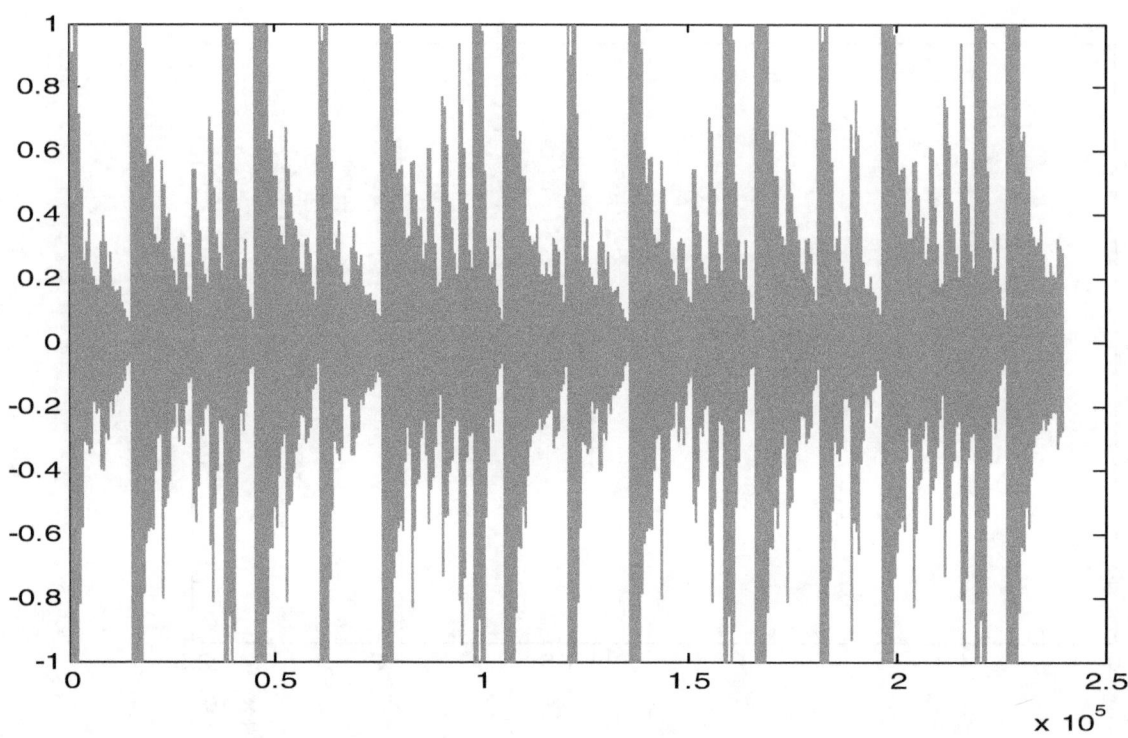

**HALF THE VOLUME OF MUSIC SIGNAL**

```
>> N= 240000;
>> [x,fs,bits]=wavread('C:\Program Files\MATLAB\R2013a\bin\Bass\signal1.wav',N);
>> wavwrite(x/2,fs,bits,'C:\Program Files\MATLAB\R2013a\bin\Bass\halfvolume.wav');
>> [HVOL,fs,bits]=wavread('C:\Program Files\MATLAB\R2013a\bin\Bass\halfvolume.wav',N);
>> wavplay(HVOL,fs);
```

PLOTING HALF VOLUME SIGNAL

plot(HVOL,'c')

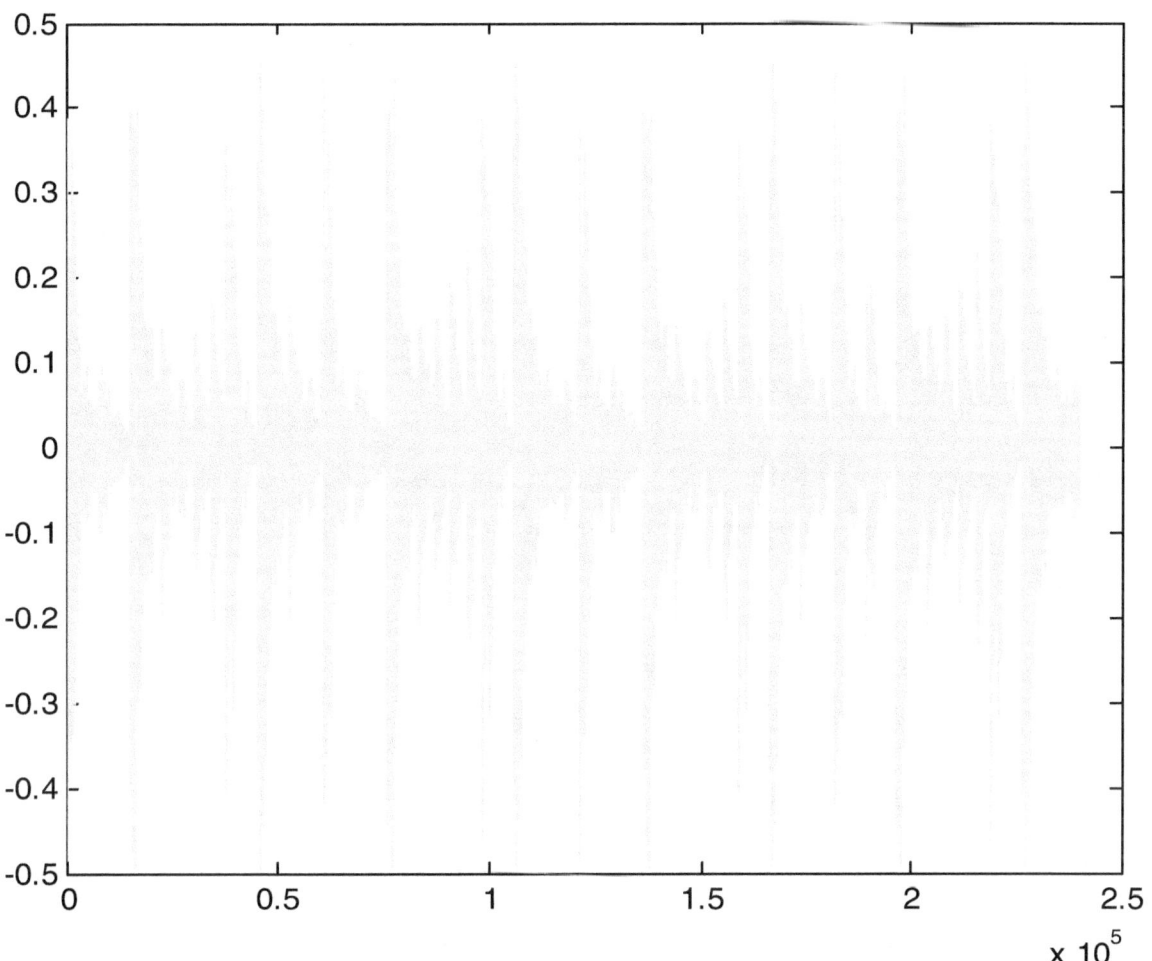

SEPERATING THE DIFFERENT FREQUENCY COFFICIENT OF AN AUDIO SIGNAL:
     AS WE KNOW THAT A MUSIC SIGNAL IS COMPOSED OF DIFFERENT FREQUENCY COMPONENTS. EVERY FREQUENCY BAND HAS A DIFFERENT SOUND CHARACTERISTICS AND HAVE A DIFFERENT FREQUENCY RANGE.
FOR EXAMPLE  SOUND OF DRUM HAVE LOW FREQUENCIES WHEREAS THE SOUND OF VIOLIN OR GUITAR HAS HIGHER FREQUENCIES. OUR EAR CAN DISCRIMINATES BETWEEN DIFFERENT FREQUENCY COMPONENTS AND WE CAN SEPARATE DIFFERENT FREQUENCY COMPONENTS OF AUDIO FREQUENCY BANDWIDTH BY PASSING THESE AUDIO SIGNALS THROUGH HIGH PASS FILTER AND LOW PASS FILTERS OF DIFFERENT CUT OFF FREQUENCIES.
 WAVELET TRANSFORM is a technique by which we can separate different frequency components of any signal.

PROCESS OF TAKING WAVELET TRANSFORM :

In the process of wavelet transform the original signal(S) is first decompose into Approximate Coefficients and Detailed Coefficients by simply passing the signal through low pass filter and high pass filter respectively.

The output of **low pass filter** is called Approximate [A1] (Low frequency components) coefficient of the signal

The output of **High pass filter** is called Detailed [D1] (High frequency components) Coefficients Of the signal.

This Approximate coefficient [A1] again passed through a low pass and high passes filter and again Decompose the signal into Approximate [A2] and Detailed Coefficients [D2].

Further Approximate Components [A2] can be decomposed into approximate coefficients [A3] And Detailed Coefficients [D3]

The number of Decomposition levels depends on the length of signal and our requirements

S   = A1+D1      [First level Wavelet Decomposition]……...…………………………..**[1]**

A1 = A2+D2      [Second level Wavelet Decomposition]…….............. **[2]**

A2 = A3+D3      [Third level Wavelet Decomposition]…….............. **[3]**


S   = A3+D3+D2+D1                        ……………………………….. **[4]**

The Original Signal S can be reconstructing with the help of A3, D3, D2 and D1.

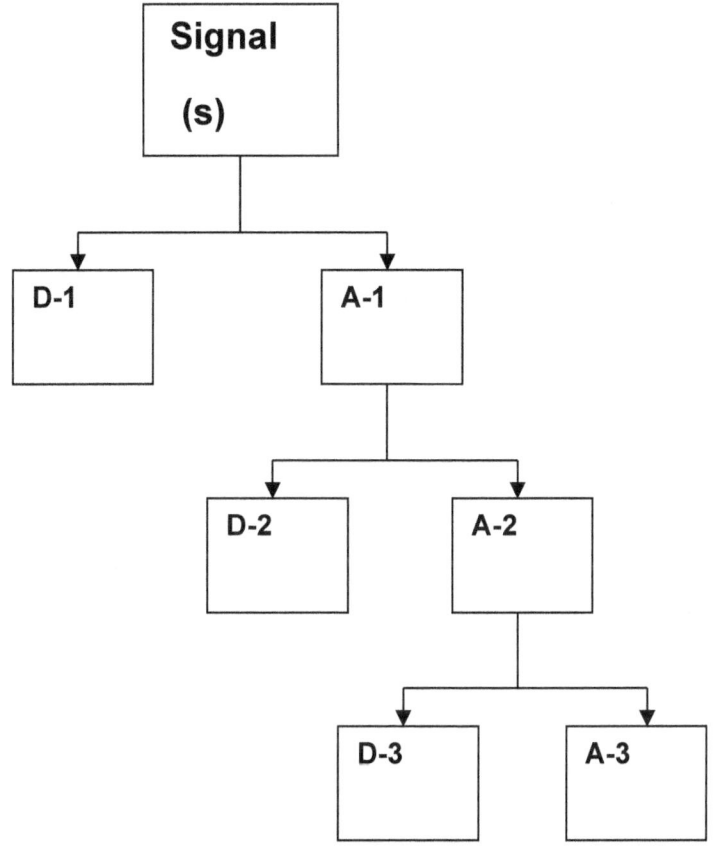

**Fig  Wavelet Decomposition of signal S=A3+D3+D2+D1**

So it is clear that with the help of equation (1) and equation (2), (3), (4)

We can decompose any original signal sequences in to wavelet decomposition. And with these wavelet decomposition again we can construct the original signal.

The number of sample in next decomposition level is half as compared to previous stage. Suppose The original signal S has N samples then A1 and D1 will have N/2 Samples and A2 and D2 will have N/4 Samples.

**SEPERATING THE DIFFERENT FREQUNCY COMPONANTS OF AN AUDIO SIGNAL**

```
N= 240000;
[x,fs,bits]=wavread('C:\Program Files\MATLAB\R2013a\bin\Bass\signal1.wav',N);
wavplay(x,fs);
pause;
```

In above matlab example wavplay is the matlab function which is used to play back audio file (.wev extension audio files) at a rate fs(44100 ) samples per second

**PLOTING AN ORIGINAL AUDIO SIGNAL**
plot(x)

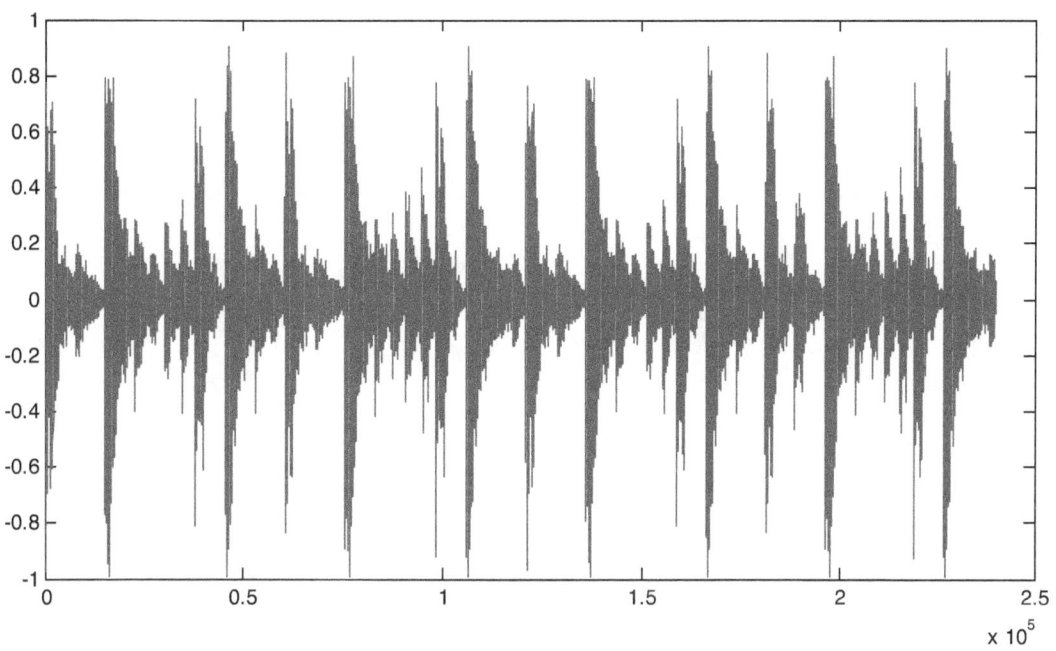

MATLAB PROGRAAM FOR SEPERATING DIFFERENT FREQUNCY COMPONANTS

**Using Wavlet transform we can separate the different frequency components from each other. To take wavlet transform of an audio signal first we need to choose any one wavlet out of several available wavlet families and again need to choose the level of decomposition.**

```
Wname = 'Haar';
N= 240000;
[x,fs,bits]=wavread('C:\Program Files\MATLAB\R2013a\bin\Bass\signal1.wav',N);

[A1,D1] = dwt(x,wname);          %first level of wavelet cofficients
                                 A1 = 1st level approximate cofficient
                                 D1 =1st level detailed cofficients
[A2,D2] = dwt(A1,wname);         %second level of wavelet cofficient
[A3,D3] = dwt(A2,wname);         % Third level of Wavelet coefficient

wavwrite(A1,fs,bits,'C:\Program Files\MATLAB\R2013a\bin\Bass\approximate1.wav');
 wavwrite(A2,fs,bits,'C:\Program Files\MATLAB\R2013a\bin\Bass\approximate2.wav');
 wavwrite(A3,fs,bits,'C:\Program Files\MATLAB\R2013a\bin\Bass\approximate3.wav');

 wavwrite(D1,fs,bits,'C:\Program Files\MATLAB\R2013a\bin\Bass\detail1.wav');
 wavwrite(D2,fs,bits,'C:\Program Files\MATLAB\R2013a\bin\Bass\detail2.wav');
```

```
wavwrite(D3,fs,bits,'C:\Program Files\MATLAB\R2013a\bin\Bass\detail3.wav');
```

In abave program we choose 'HAAR' Wavlet number of sample of an audio signal is 240000
Dwt is the discreate wavelet function which takes the wavlet transform of signal x and
store first approximate coefficients to the variable A1 and store first detail coefficient
to variable D1

Again these first approximate coefficient[A1] is further decomposed into A2 and D2 and
again these second approximate coefficients [A2] is further decomposed into A3 and D3.

**FIRST APPROXIMATE [LOW FREQUENCY COFFICIENT] COFFICIENTS:**
```
Wname = 'Haar';
N= 240000;
[x,fs,bits]=wavread('C:\Program Files\MATLAB\R2013a\bin\Bass\signal1.wav',N);

[A1,D1] = dwt(x,wname);
```

```
wavwrite(A1,fs,bits,'C:\Program Files\MATLAB\R2013a\bin\Bass\approximate1.wav');

wavplay(A1,fs);
plot(A1);
pause;
```

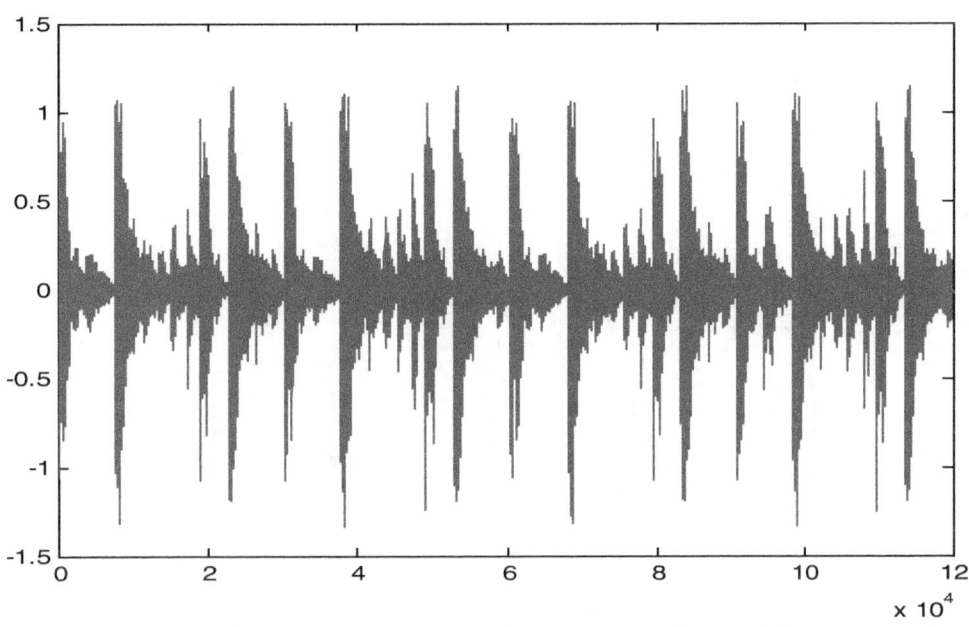

As we know that A1 has number of sample that is half of original signal x. So the length of
A1 is half of signal x. So we should play A1 at a half rate of that we played x.

```
wavplay(A1,fs/2);
```

**SECOND APPROXIMATE [LOW FREQUENCY COFFICIENT] COFFICIENTS:**
```
Wname = 'Haar';
N= 240000;
[x,fs,bits]=wavread('C:\Program Files\MATLAB\R2013a\bin\Bass\signal1.wav',N);

[A1,D1] = dwt(x,wname);         %first level of wavelet cofficients
                                A1 = 1st level approximate cofficient
```

```
                                    D1 =1st level detailed cofficients
[A2,D2] = dwt(A1,wname);            %second level of wavelet cofficient

wavwrite(A2,fs,bits,'C:\Program Files\MATLAB\R2013a\bin\Bass\approximate2.wav');

wavplay(A2,fs);

plot(A2);
pause;
```

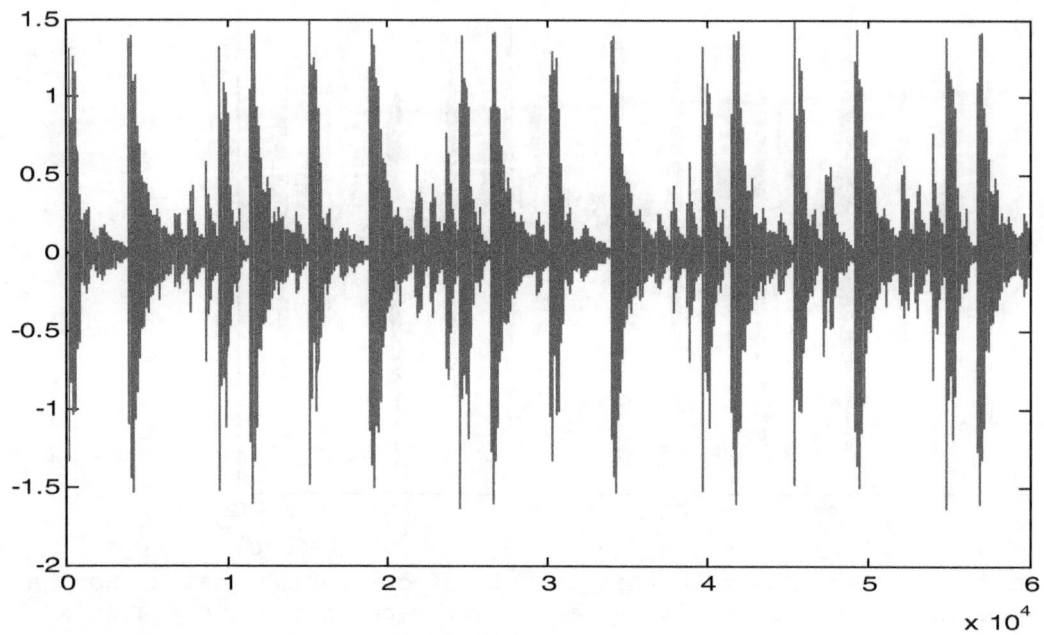

A2 has the number of sample that is again half of A1. And one fourth of x so
we should play A2 at a rate that is one fourth of rate we played x.
```
wavplay(A2,fs/4);
```

**THIRD APPROXIMATE [LOW FREQUENCY COFFICIENT] COFFICIENTS:**
```
Wname = 'Haar';
N= 240000;
[x,fs,bits]=wavread('C:\Program Files\MATLAB\R2013a\bin\Bass\signal1.wav',N);

[A1,D1] = dwt(x,wname);             %first level of wavelet cofficients
                                    A1 = 1st level approximate cofficient
```

```
                                    D1 =1st level detailed cofficients
[A2,D2] - dwt(A1,wname);            %second level of wavelet cofficient
[A3,D3] = dwt(A2,wname);            % Third level of Wavelet coefficient


wavplay(A3,fs);

wavwrite(A3,fs,bits,'C:\Program Files\MATLAB\R2013a\bin\Bass\approximate3.wav');

plot (A3);
pause;
```

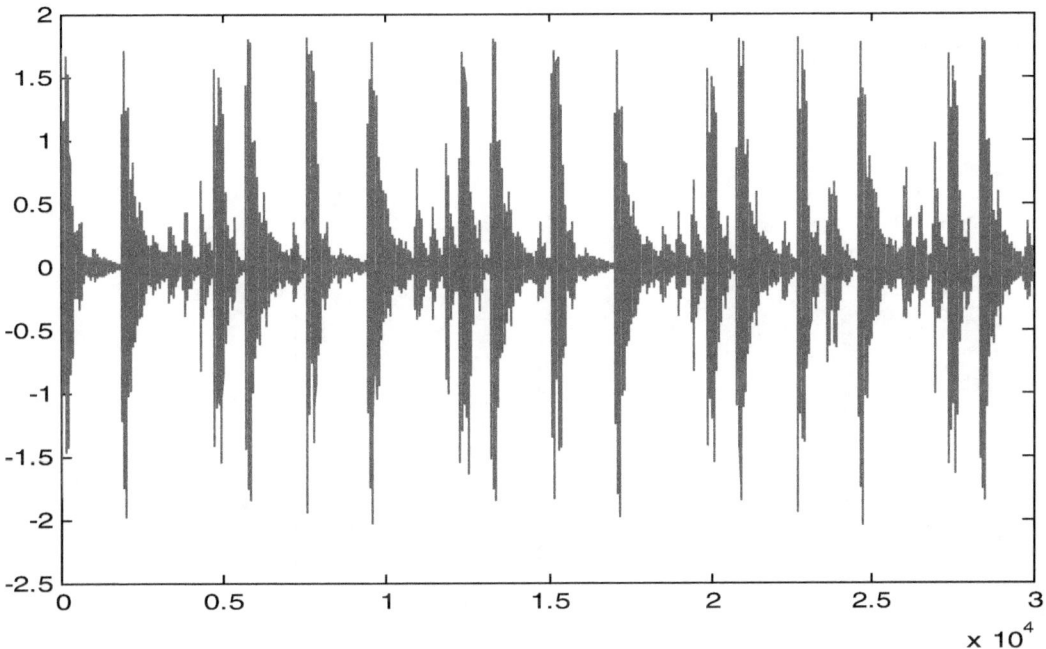

A3 has the number of sample that is half of A2. And one eight of original signal x. so we should play A3 at a rate one eight of that we played x.

```
wavplay(A3,fs/8);
```

**FIRST DETAIL [HIGH FREQUENCY COFFICIENT] COFFICIENTS:**
```
Wname = 'Haar';
N= 240000;
[x,fs,bits]=wavread('C:\Program Files\MATLAB\R2013a\bin\Bass\signal1.wav',N);

[A1,D1] = dwt(x,wname);             %first level of wavelet cofficients
                                    A1 = 1st level approximate cofficient
```

```
                                     D1 =1st level detailed cofficients
[A2,D2] = dwt(A1,wname);             %second level of wavelet cofficient
[A3,D3] = dwt(A2,wname);             % Third level of Wavelet coefficient

 wavwrite(D1,fs,bits,'C:\Program Files\MATLAB\R2013a\bin\Bass\detail1.wav');
 wavwrite(D2,fs,bits,'C:\Program Files\MATLAB\R2013a\bin\Bass\detail2.wav');
 wavwrite(D3,fs,bits,'C:\Program Files\MATLAB\R2013a\bin\Bass\detail3.wav');


wavplay(D1,fs);
pause;
plot(D1,'g')
```

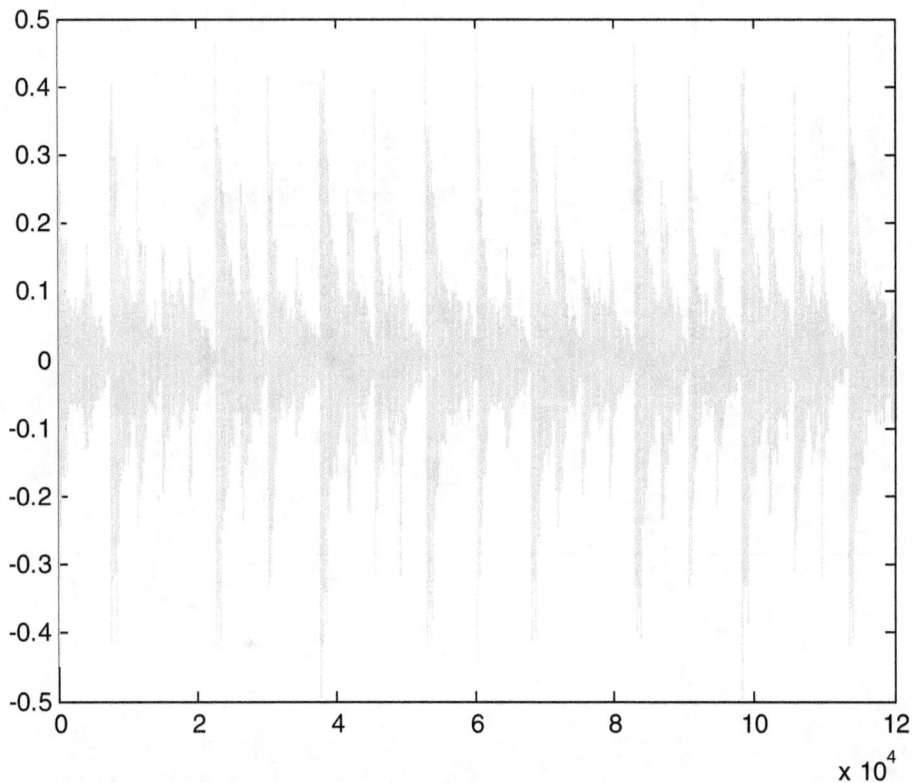

D1 has the number of sample that is half of original signal X. So we should
play D1 at a rate twice that of we played original signal x.


```
SECOND DETAIL [HIGH FREQUENCY COFFICIENT] COFFICIENTS:
Wname = 'Haar';
N= 240000;
[x,fs,bits]=wavread('C:\Program Files\MATLAB\R2013a\bin\Bass\signal1.wav',N);

[A1,D1] = dwt(x,wname);              %first level of wavelet cofficients
                                     A1 = 1st level approximate cofficient
```

```
                              D1 =1st level detailed cofficients
[A2,D2] = dwt(A1,wname);      %second level of wavelet cofficient
[A3,D3] = dwt(A2,wname);      % Third level of Wavelet cocfficient

  wavwrite(D2,fs,bits,'C:\Program Files\MATLAB\R2013a\bin\Bass\detail2.wav');


wavplay(D2,fs);
pause;
plot(D2,'g')
```

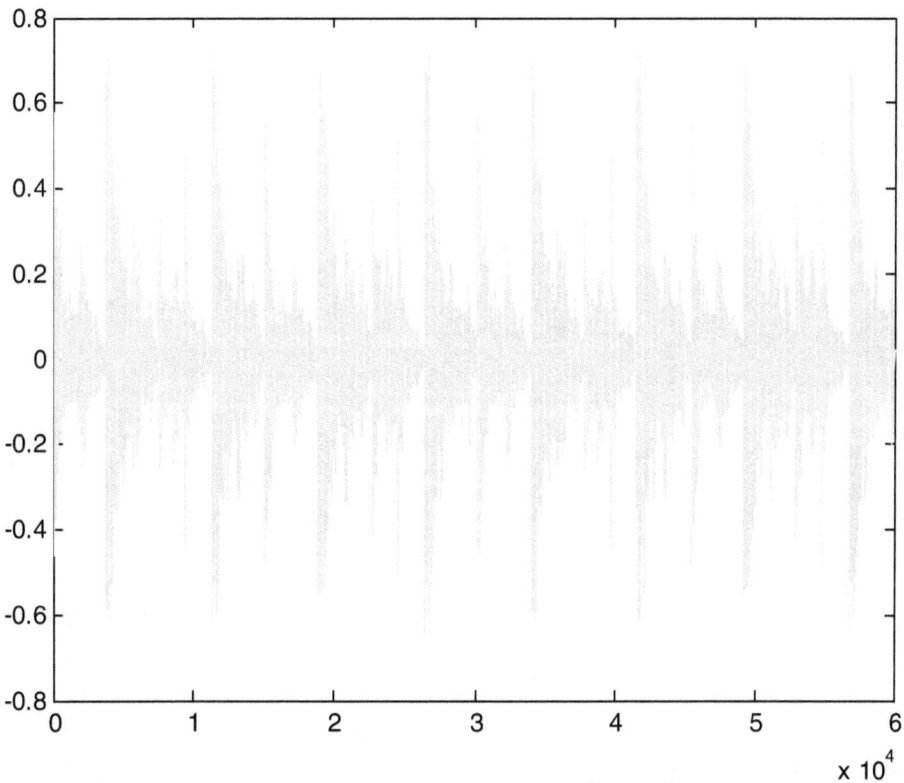

D2 has the number of samples that is half of D1 and one fourth of original
signal x. so we should play D2 at a rate that is one fourth we played x.


**THIRD DETAIL [HIGH FREQUENCY COFFICIENT] COFFICIENTS:**
```
Wname = 'Haar';
N= 240000;
[x,fs,bits]=wavread('C:\Program Files\MATLAB\R2013a\bin\Bass\signall.wav',N);

[A1,D1] = dwt(x,wname);          %first level of wavelet cofficients
                                 A1 = 1st level approximate cofficient
```

```
                              D1 =1st level detailed cofficients
[A2,D2] = dwt(A1,wname);      %second level of wavelet cofficient
[A3,D3] = dwt(A2,wname);      % Third level of Wavelet coefficient

wavwrite(D3,fs,bits,'C:\Program Files\MATLAB\R2013a\bin\Bass\detail3.wav');

wavplay(D3,fs);
pause;
plot(D3,'g')
```

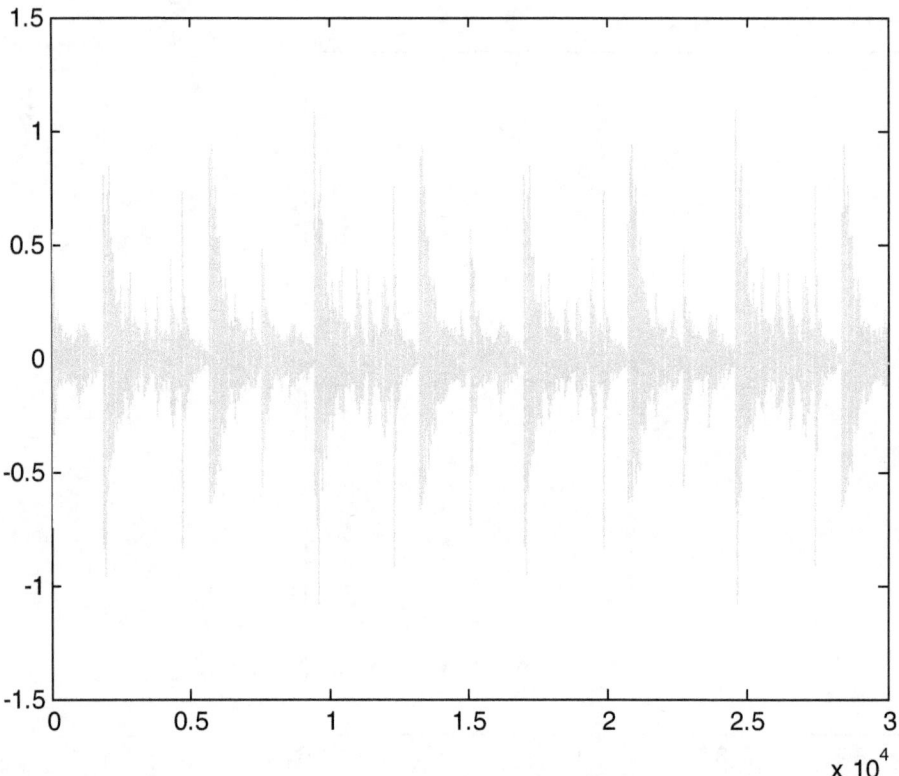

D3 has the number of sample that is half of D2 And one eight of original signal x. so we should play D3 at a rate one eight of that we played original signal x.

```
wavplay(D3,fs/8);
```

## AMPLIFYING D3 COMPONANT OF AN AUDIO SIGNAL

```
Wname = 'Haar';
N= 240000;
[x,fs,bits]=wavread('C:\Program Files\MATLAB\R2013a\bin\Bass\signal1.wav',N);
```

```
[A1,D1] = dwt(x,wname);              %first level of wavelet cofficients
                                     A1 = 1st level approximate cofficient
                                     D1 =1st level detailed cofficients
[A2,D2] = dwt(A1,wname);             %second level of wavelet cofficient
[A3,D3] = dwt(A2,wname);             % Third level of Wavelet coefficient


A22=idwt(A3,D3*20,wname);
A11=idwt(A22,D2,wname);
x1=idwt(A11,D1,wname);


A23=idwt(A3,D3*30,wname);
A13=idwt(A23,D2 ,wname);
x2=idwt(A13,D1,wname);


A24=idwt(A3,D3*40,wname);
A14=idwt(A24,D2,wname);
x3=idwt(A14,D1,wname);


wavplay(x1,fs);
pause;
plot(x1,'r');
wavwrite(x1,fs,bits,'C:\Program Files\MATLAB\R2013a\bin\Bass\amplifydetail1.wav');
```

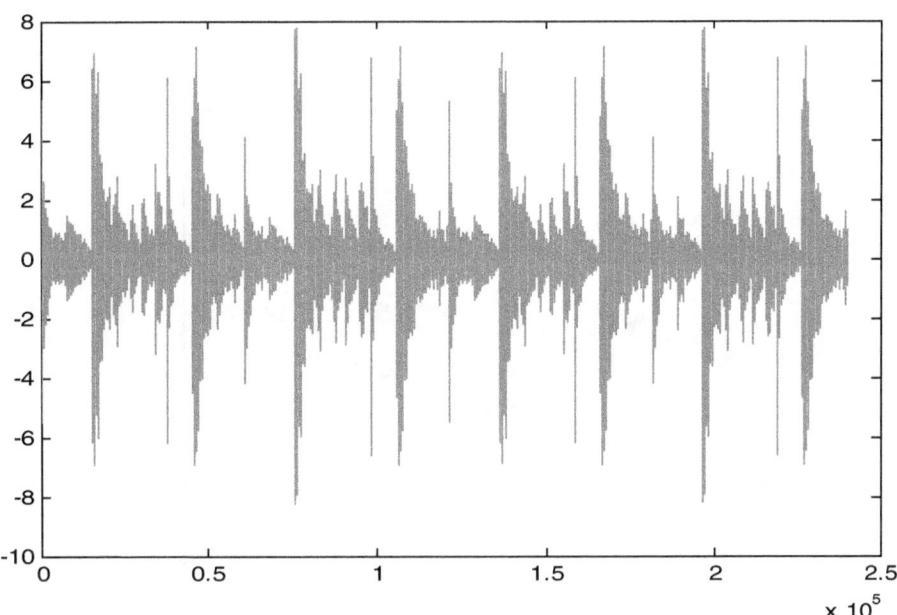

```
wavplay(x2,fs);
pause;
plot(x2,'y');
wavwrite(x2,fs,bits,'C:\Program Files\MATLAB\R2013a\bin\Bass\amplifydetail2.wav');
```

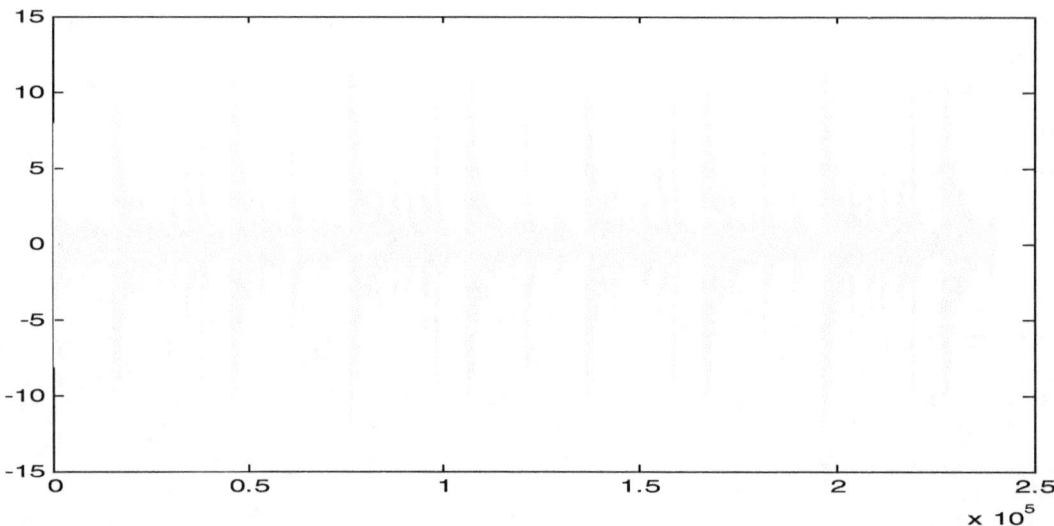

```
wavplay(x3,fs);
pause;
plot(x3,'c');
wavwrite(x3,fs,bits,'C:\Program Files\MATLAB\R2013a\bin\Bass\amplifydetail3.wav');
```

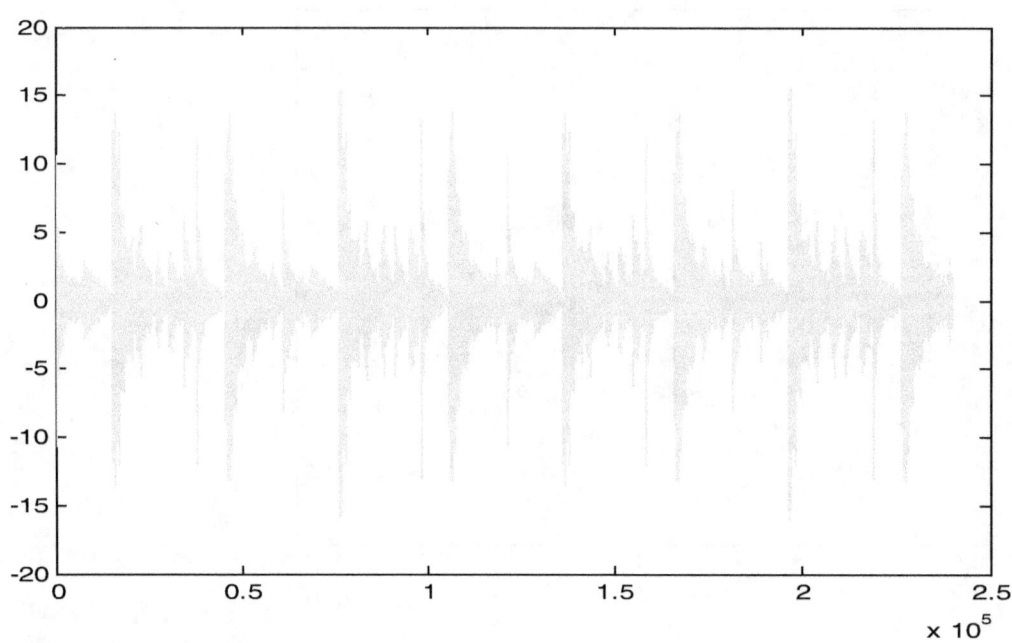

AMPLIFYING D1 [FIRST DETAIL] COMPONANT OF AN AUDIO SIGNAL

[1] TEN TIMES OF ORIGINAL SOUND

```
wname = 'Haar';
N= 240000;
[x,fs,bits]=wavread('C:\Program
Files\MATLAB\R2013a\bin\Bass\signal1.wav',N);

[A1,D1] = dwt(x,wname);
x1=idwt(A1,D1*10,wname);
wavplay(x1,fs);
pause;
plot(x1,'r');
wavwrite(x1,fs,bits,'C:\Program
Files\MATLAB\R2013a\bin\Bass\amplifyfirstdetail10.wav');
```

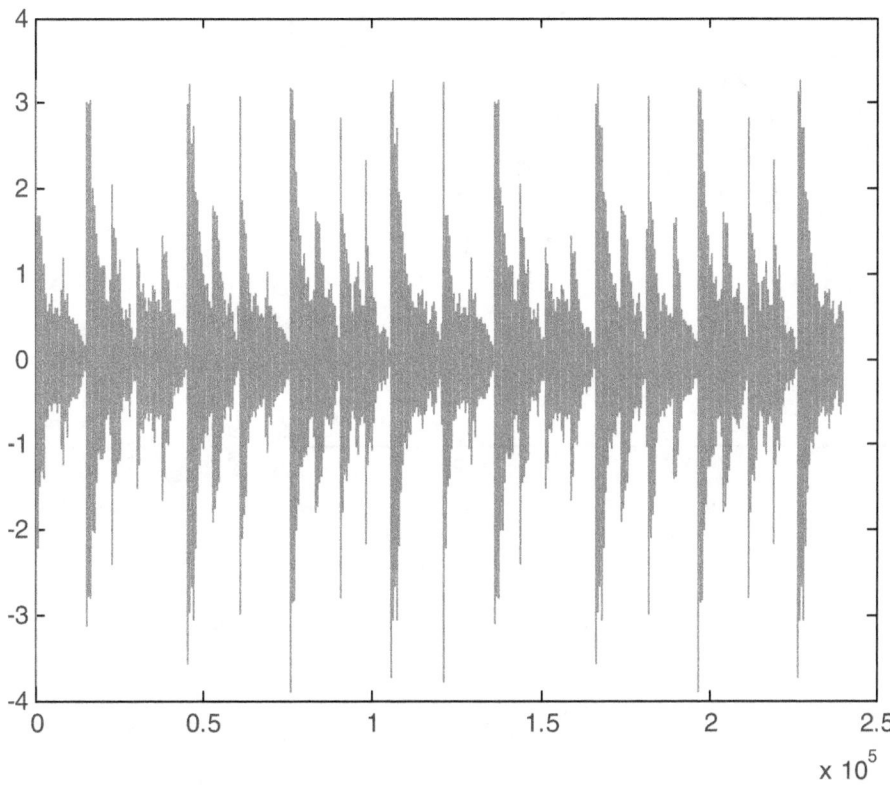

Wave Sound

## [2] TWENTY TIMES OF ORIGINAL SOUND

```
>> wname = 'Haar';
N= 240000;
[x,fs,bits]=wavread('C:\Program Files\MATLAB\R2013a\bin\Bass\signal1.wav',N);
[A1,D1] = dwt(x,wname);
```

```
x1=idwt(A1,D1*20,wname);
wavplay(x1,fs);
pause;
plot(x1);
wavwrite(x1,fs,bits,'C:\Program Files\MATLAB\R2013a\bin\Bass\amplifyfirstdetail20.wav');
```

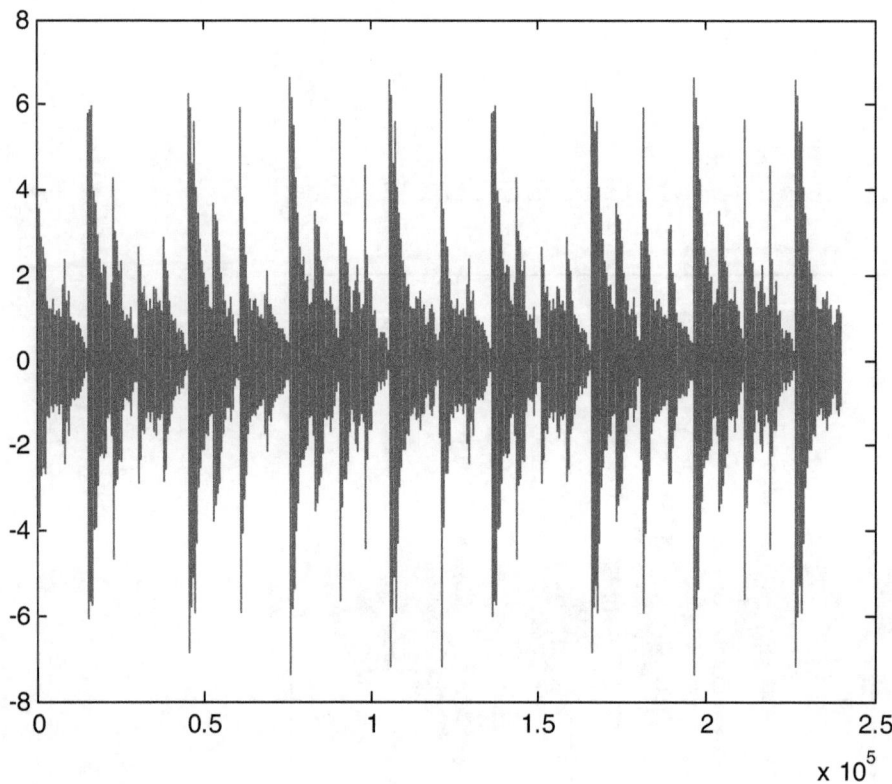

[2] THIRTY TIMES OF ORIGINAL SOUND

```
wname = 'Haar';
N= 240000;
[x,fs,bits]=wavread('C:\Program Files\MATLAB\R2013a\bin\Bass\signal1.wav',N);
```

```
[A1,D1] = dwt(x,wname);
x1=idwt(A1,D1*30,wname);
wavplay(x1,fs);
pause;
plot(x1,'G');
wavwrite(x1,fs,bits,'C:\Program Files\MATLAB\R2013a\bin\Bass\amplifyfirstdetail30.wav');
```

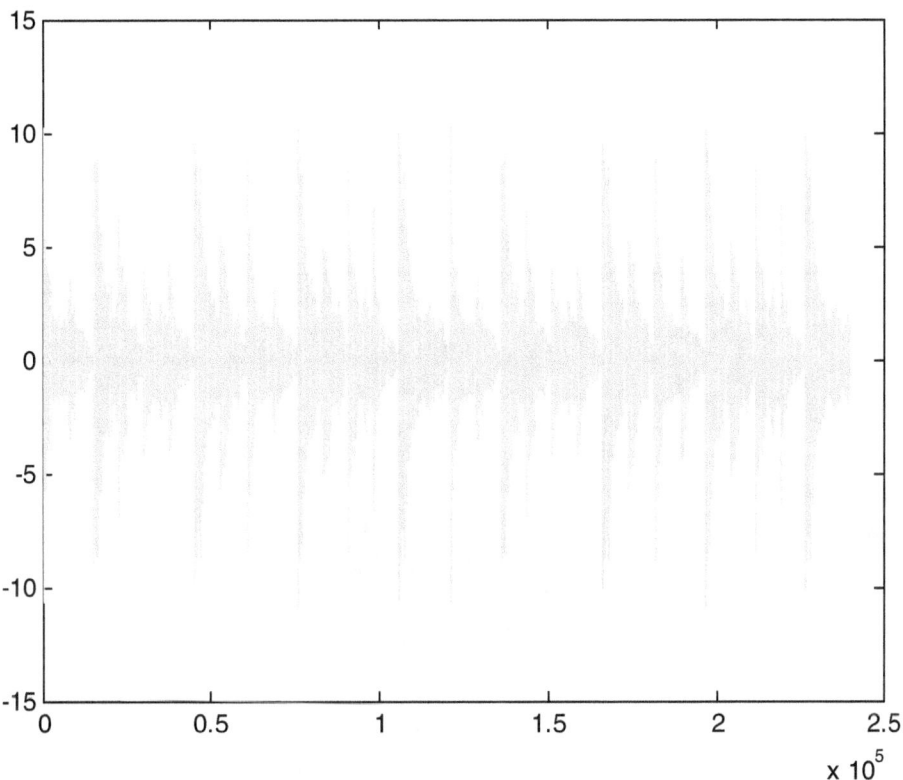

AMPLIFYING A1 [FIRST APPROXIMATE] COMPONANT OF AN AUDIO SIGNAL

[1] TEN TIMES OF ORIGINAL SOUND

```
>> wname = 'Haar';
```

```
N= 240000;
[x,fs,bits]=wavread('C:\Program
Files\MATLAB\R2013a\bin\Bass\signal1.wav',N);

[A1,D1] = dwt(x,wname);
x1=idwt(A1*10,D1,wname);
wavplay(x1,fs);
pause;
plot(x1,'r');
wavwrite(x1,fs,bits,'C:\Program
Files\MATLAB\R2013a\bin\Bass\amplifyfirstapproximate10.wav');
```

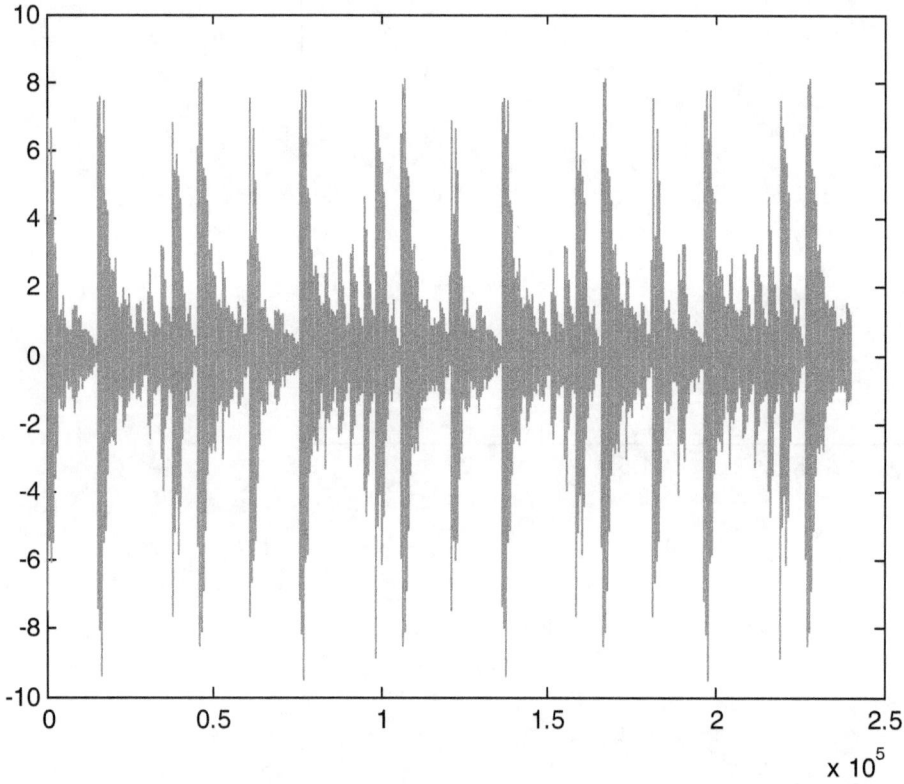

[2] TWENTY TIMES OF ORIGINAL SOUND

```
>> wname = 'Haar';
```

```
N= 240000;
[x,fs,bits]=wavread('C:\Program Files\MATLAB\R2013a\bin\Bass\signal1.wav',N);
[A1,D1] = dwt(x,wname);
x1=idwt(A1*20,D1,wname);
wavplay(x1,fs);
pause;
plot(x1,'r');
wavwrite(x1,fs,bits,'C:\Program
Files\MATLAB\R2013a\bin\Bass\amplifyfirstapproximate20.wav');
```

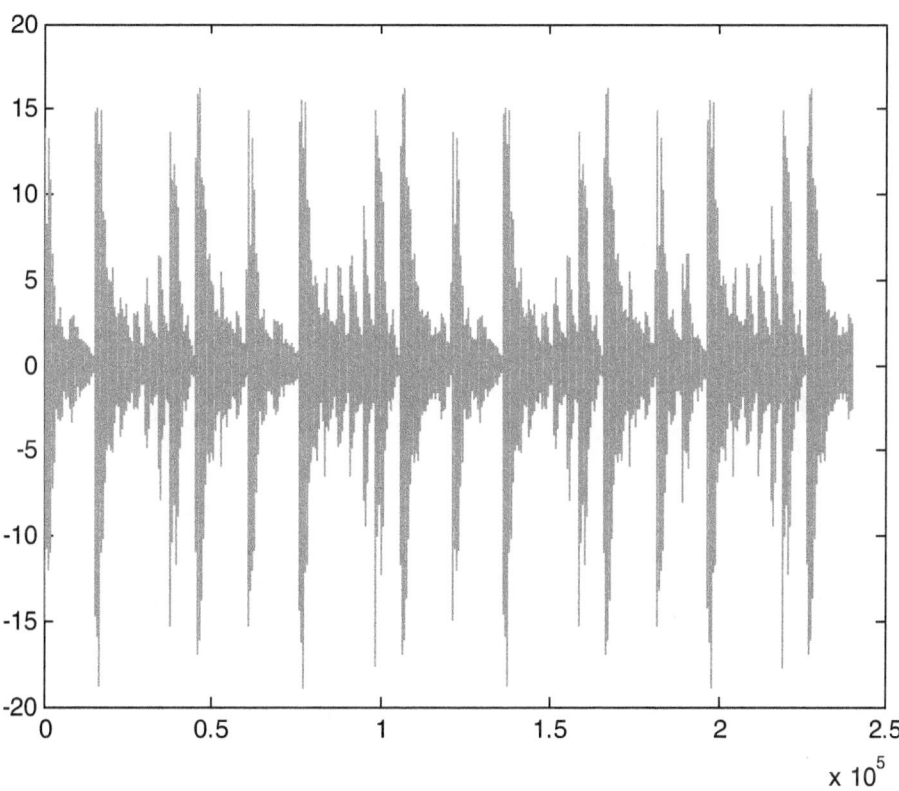

[2] THIRTY TIMES OF ORIGINAL SOUND

```
>> wname = 'Haar';
N= 240000;
```

```
[x,fs,bits]=wavread('C:\Program Files\MATLAB\R2013a\bin\Bass\signal1.wav',N);
[A1,D1] = dwt(x,wname);
x1=idwt(A1*30,D1,wname);
wavplay(x1,fs);
pause;
plot(x1,'G');
wavwrite(x1,fs,bits,'C:\Program
Files\MATLAB\R2013a\bin\Bass\amplifyfirstapproximate30.wav');
```

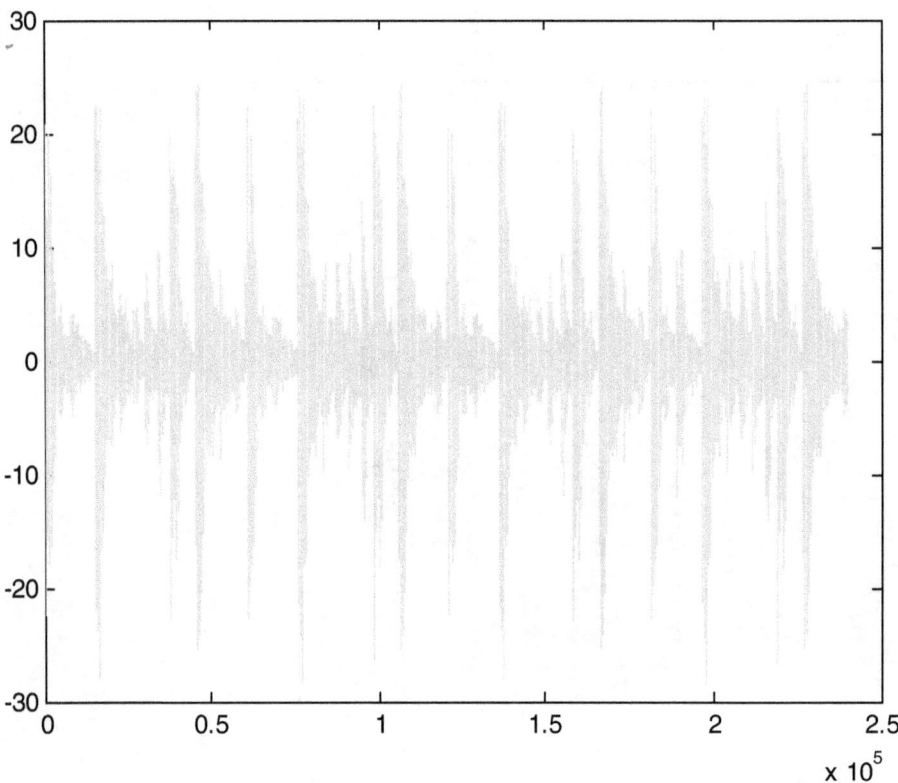

DIFFERENT VARIABLES USED IN MATLAB PROGRAMMING

A1,    <120000x1 double>

```
A11,   <120000x1 double>
A13,   <120000x1 double>
A14,   <120000x1 double>
A2,    <60000x1 double>
A22,   <60000x1 double>
A23,   <60000x1 double>
A24,   <60000x1 double>
A3,    <30000x1 double>
D1,    <120000x1 double>
D2,    <60000x1 double>
D3,    <30000x1 double>
N,     240000
bits,  16,
fs,    44100,
wname, 'haar',
x,       <240000x1 double>
x1,      <240000x1 double>
x2,      <240000x1 double>
x3       <240000x1 double>
```

## ADDING TWO OR MORE MUSIC FILE

```
N= 240000;

[x,fs,bits]=wavread('C:\Program Files\MATLAB\R2013a\bin\Bass\signal1.wav',N);
```

```matlab
[y,fs,bits]=wavread('C:\Program Files\MATLAB\R2013a\bin\GUITTAR\opentune.wav',N);

[z,fs,bits]=wavread('C:\Program Files\MATLAB\R2013a\bin\GUITTAR\S5_f00.wav',N);

Music = x+y+z;

wavwrite(x1,fs,bits,'C:\Program Files\MATLAB\R2013a\bin\Bass\Music');

C:\Program Files\MATLAB\R2013a\bin\GUITTAR


wavplay(x,fs);
pause;
plot(x,'r');

wavplay(y,fs);
pause;
plot(y,'g');

wavplay(z,fs);
pause;
plot(x,'b');


wavplay(music,fs);
pause;
plot(music,'c');
```

## ADDING TWO OR MORE MUSIC FILE

```matlab
N= 240000;
```

```
[x,fs,bits]=wavread('C:\Program Files\MATLAB\R2013a\bin\Bass\signal1.wav',N);
[y,fs,bits]=wavread('C:\Program Files\MATLAB\R2013a\bin\GUITTAR\opentune.wav',N);
[z,fs,bits]=wavread('C:\Program Files\MATLAB\R2013a\bin\CUITTAR\S5_f00.wav',N);

Music = x+y+z;

wavplay(x,fs);
```

plot(x,'r');

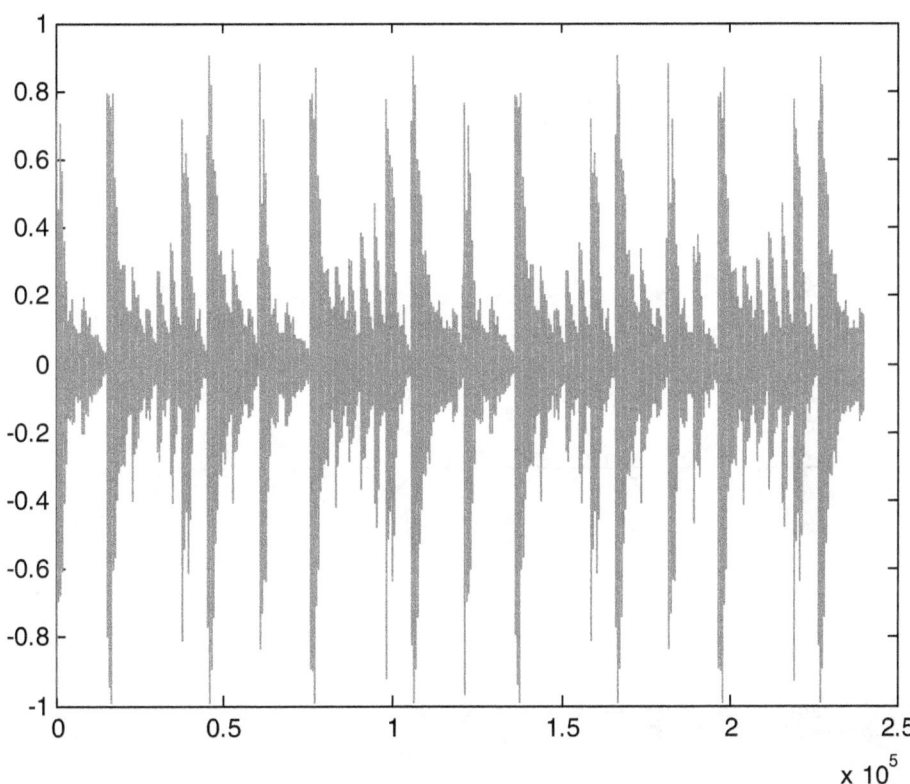

```
wavplay(y,fs);
```

plot(y,'g');

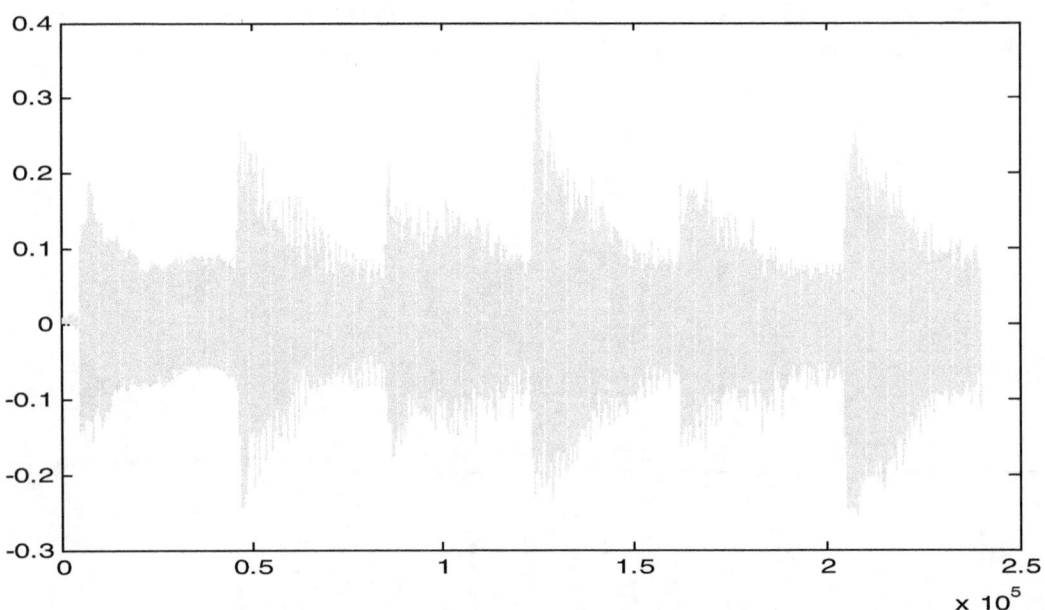

```
wavplay(z,fs);
plot(z)
```

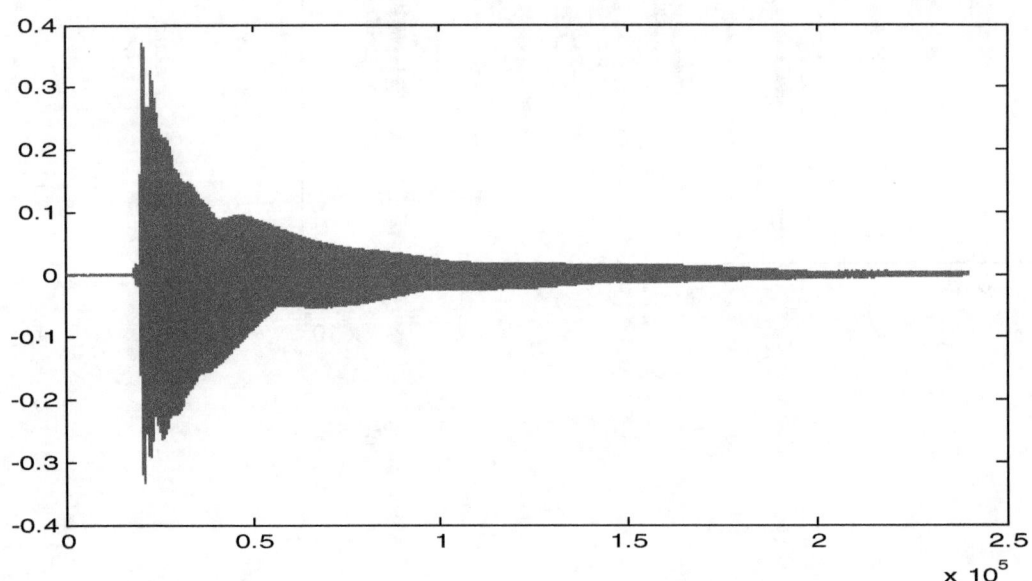

```
wavplay(Music,fs);
plot(music,'c');
```

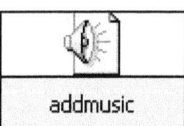
addmusic

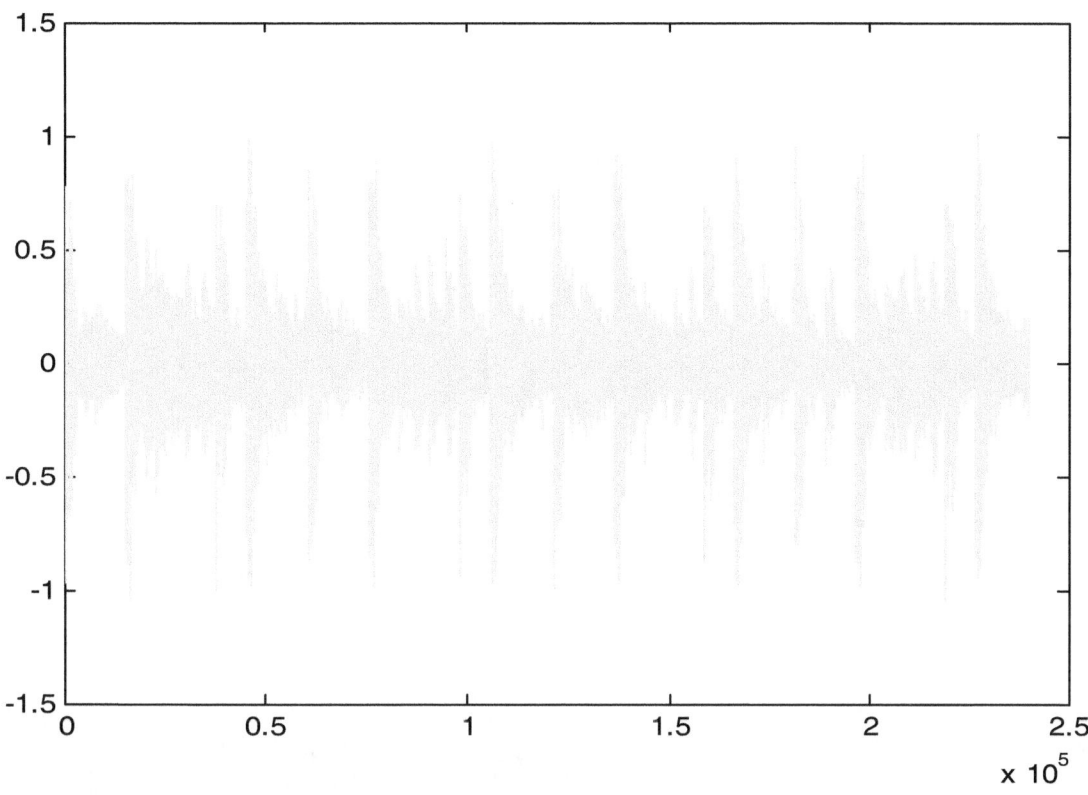

## THANKS

Waiting for feedback and suggestions at
vishnunarayan.saxena@gmail.com

www.ingramcontent.com/pod-product-compliance
Lightning Source LLC
Chambersburg PA
CBHW080609190526
45169CB00007B/2942